Black Bear Dreams

Anne Michelle Lawrence

Illustrated by Ken Silbert

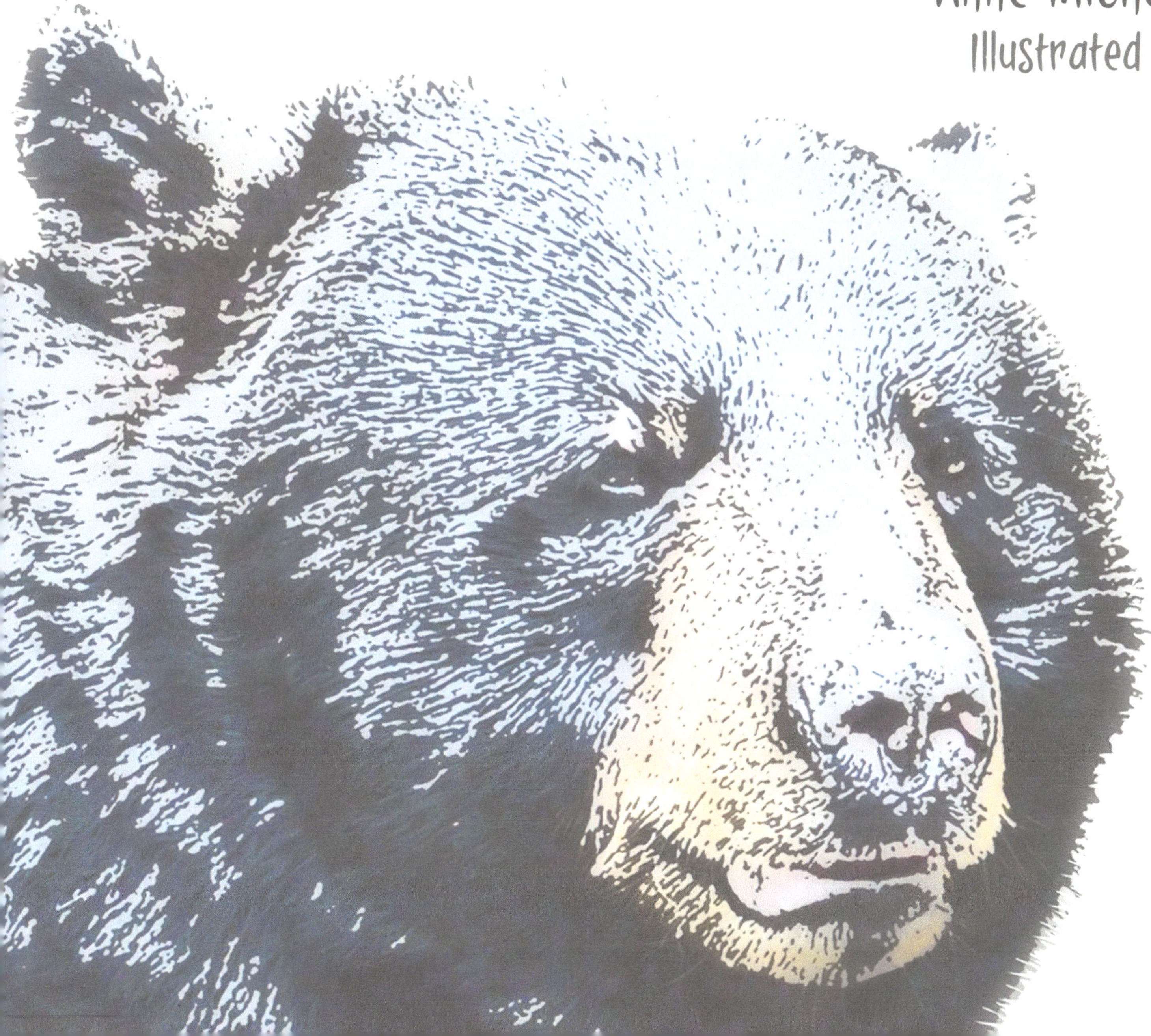

Black Bear Dreams
Copyright © 2020 by Anne Michelle Lawrence

All rights reserved. No part of this publication may be reproduced in any form or by any electronic or mechanical means, including information storage and retrieval systems, without the express written permission from the publisher, except in the case of brief quotations embodied in critical articles or reviews. For information regarding permission, contact BeaLuBooks.

ISBN: 978-0-9990924-5-3

Library of Congress Control Number: 2019952451
Publisher's Cataloging-in-Publication Data is on file with the publisher.

Edited by: Luana K. Mitten
Book cover and interior design by Ken Silbert

Printed in the United States of America
December 2019

BeaLu Books
Tampa, Florida

www.BeaLuBooks.com

Photo Credits:
Front Cover: (bear)* Greg Hume; Back Cover: (bear)* Lorraine Logan/Shutterstock; page 4*: CCo Creative Commons; page 5*: Menno Schaefer/Shutterstock; page 6: (classroom)* Vectorpocket/Shutterstock, (bear) Orfeev/Shutterstock; page 7: (bears) Nebojsa Kontic/Shutterstock; page 8*: Hal Brindley/Shutterstock; page 10: (bear)* Lorraine Logan/Shutterstock; page 11*: Lisa Parsons/Shutterstock; page 12: (roots)* N. American Bear Center; page 13: (bears)* Orfeev/Shutterstock, (humans)* Chipmunk131/Shutterstock; page 14: (top left) Cyrustr/Shutterstock, (top middle) Alexander Pigagis/Shutterstock, (top right) Kateryna Mashkevych, (bottom) LeManna/Shutterstock; page 15*: Nick Fox/Shutterstock; page 16*: Lorraine Logan/Shutterstock; page 17: (top)* KC Cheyne/Shutterstock, (bottom right) FedBull.Shutterstock, (middle) Armando Frazao/Shutterstock; page 18: (cake)* Romiana Lee/Shutterstock, (bear)* Lorraine Logan/Shutterstock; page 20: (acorns) Kimberly Irish/Shutterstock, (leaves)* Volosina/Shutterstock, (cabbage) Ron Rowan Photography/Shutterstock, (berries) Joanna Tkaczuk/Shutterstock, (maple)* LunarVogel/Shutterstock; page 21: (larvae) Henrik Larrson/Shutterstock, (frog) Michael Benard/Shutterstock, (squirrel) Brian Lasenby/Shutterstock, (fish) CSNafzger/Shutterstock, (bear)* Andrea Izzotti/Shutterstock; page 22*: Hans.P/Shutterstock; page 23* Glass and Nature/Shutterstock; page 24: (ice cream) M. Unal Ozmen/Shutterstock, (bear)* Lorraine Logan/Shutterstock; page 25: (tree)* Jiri Vaclavek/Shutterstock, (bears)* N. American Bear Center; page 26: Evelyn D. Harrison/Shutterstock; page 27: (bear)* Susan Kehoe/Shutterstock, (icons) Voysla/Shutterstock; page 28*: N. American Bear Center; page 29*: Glass and Nature/Shutterstock; page 30: Mike Pellinni/Shutterstock; page 31: Iron May/Shutterstock; page 32: (baby)* Darq/Shutterstock; page 33*: Mark Betram/U.S. Fish and Wildlife Service; page 34*: Geoffrey Kuchera/Shutterstock; page 35*: Orfeev/Shutterstock; page 36* Jim Cumming/Shutterstock; page 37*: Jean-Edouard Rozey/Shutterstock; page 38*: Susan Kehoe/Shutterstock; page 39: (bear)* N. American Bear Center; page 40*: PhotosByAndy/Shutterstock; page 41*: Susan Kehoe/Shutterstock; page 42*: PhotosByAndy/Shutterstock; page 43: (bear)* MR Silophop Pongsai/Shutterstock; page 44: (bear)* ODNR Division of Wildlife, (beehive) Jonathan Silbert; page 45: (bear)* ODNR Division of Wildlife, (soccer field) CEW/Shutterstock; page 46: (bear)* ODNR Division of Wildlife, (pool) Trong Nguyen/Shutterstock; page 47*: K Quinn Ferris/Shutterstock; page 48*: Svetlana Foote/Shutterstock; page 49*: Scenic Shutterbug/Shutterstock; page 50: (bear)* Lorraine Logan/Shutterstock
*Photo altered.

Table of Contents

Black Bear Country 4-9
Spring 10-15
Summer 16-21
Autumn 22-29
Winter 30-35
A New Year 36-47
Five Senses of Black Bears 44-48
Questions for Discussion 49
Glossary 50

BLACK BEAR COUNTRY

The scientific name for the Black Bear, the most common bear in North America, is *Ursus americanus*, or "American Bear."

Black bear country stretches across the continent of North America, as far north as the frozen lakes of Canada, as far south as the desert shrubs of Mexico, as far west as the rain-soaked firs of Oregon, and as far east as the rocky shores of Maine.

FACT:

The yellow on this map shows the current territory inhabited by black bears. Do you live in black bear country?

However, the average home range for an adult female black bear is about 20 square miles—roughly the size of a half million elementary classrooms pressed together.

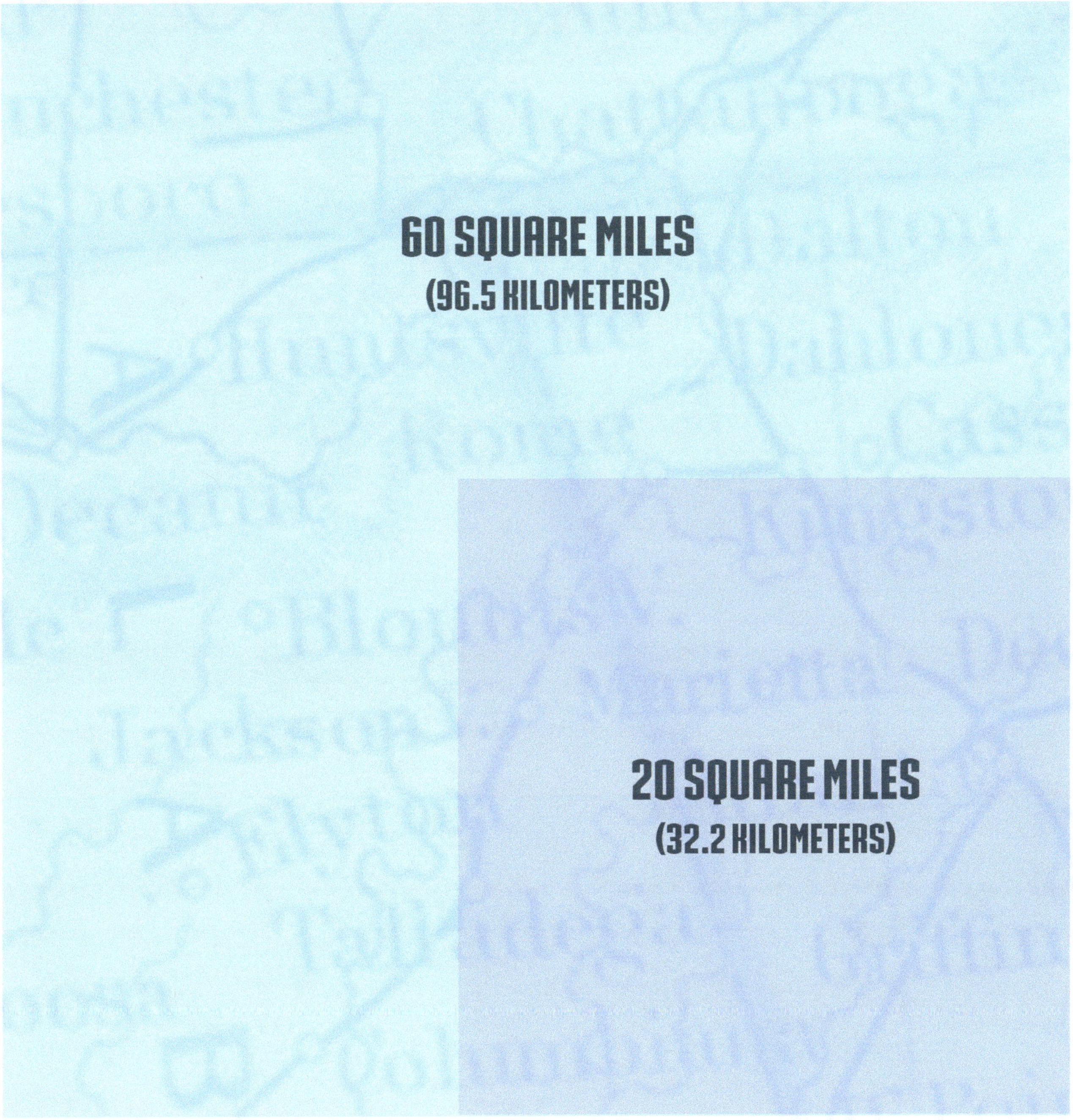

60 SQUARE MILES
(96.5 KILOMETERS)

20 SQUARE MILES
(32.2 KILOMETERS)

FACT:

For an adult male black bear, the average home range is three times as large as the range of a female black bear.

 MALE RANGE

 FEMALE RANGE

If black bears live undisturbed in the wilderness, they may be **diurnal**, or eating during the day and sleeping at night.

If black bears live near grizzly bears or humans, they may be **nocturnal**, or asleep during the day and active at night. Black bears are less threatened by competitors for food, water, and space at night.

SPRING

When the first spring moon rises,
clear and bright as birdsong,
Black Bear hears that it is time to leave,
to crawl from her winter den,
turning over stiff earth with her claws—
time to search for green shoots.

All through the night, Black Bear dreams of the swelling music of meadow brooks and the sweet hum of honeybees.
She remembers the laughter of joyful frogs.

Black bears **hibernate**, or take a long rest in a den during the coldest months of the year, when food is scarce.

During hibernation, adult male black bears, and females who did not give birth that winter, lose about one quarter of their body weight.

**ADULT FEMALE:
90-300 POUNDS
(41-136 KILOGRAMS)**

**ADULT MALE:
125-500 POUNDS
(57-227 KILOGRAMS)**

**FEET
(183 CENTIMETERS)**

FACT: The normal length and weight of adult black bears is similar to the range of adult human body sizes.

CLOVER

PINE NUTS

WILLOW BUDS

After leaving their dens in the spring, black bears reactivate their bodies by eating mostly greens, seeds, and buds.

DANDELIONS

Black bears roam for miles to **forage**, or to search for food. The distance traveled by black bears changes, depending on where they can safely find enough food.

SUMMER

When the summer moon rises,
high and bold as rainbows,
Black Bear tastes that it is time to grow,
to eat her fill again,
restoring her size, strength, and spirit—
time to climb apple trees.

All through the night, Black Bear dreams
of the sun-warmed juices of plump berries
and the river-chilled flesh of fish.
She remembers the flavors of forest herbs.

MACINTOSH APPLES

WILD STRAWBERRIES

RAINBOW TROUT

During the summer months, an adult female black bear may eat as many as 8,000 calories every day—the equivalent of a chocolate birthday cake—to replenish needed nutrients and fat.

Without these energy stores, she will not be able to become pregnant in June or to deliver baby cubs in her den the following winter.

Black bears are **omnivores**, though most of their diet consists of plants rather than meat.

ACORNS

DANDELION LEAVES

SKUNK CABBAGE ROOT

BLUEBERRIES

MAPLE FLOWERS

FACT: From root to fruit, black bears eat different parts of plants.

Black bears eat insects, including ants, beetles, and honeybees. They also may eat the meat of small animals, such as fish, frogs, and rodents, and larger animals, like young deer and elk.

FACT: Sometimes black bears will scavenge for the meat of dead animals, called carrion.

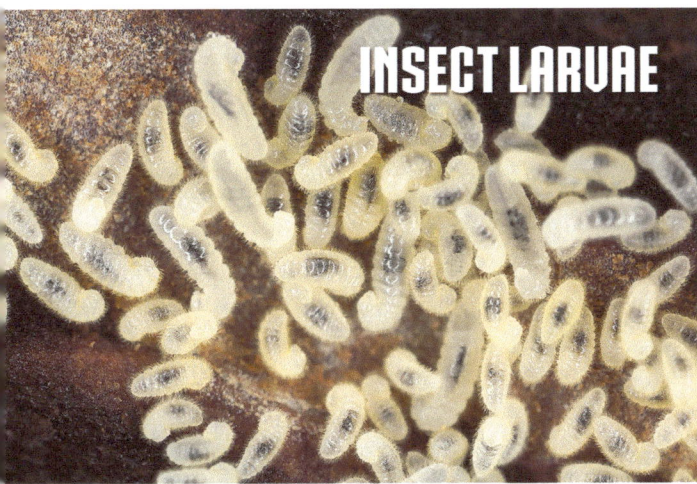
INSECT LARVAE

AMPHIBIANS

MAMMALS

FISH

AUTUMN

When the autumn moon rises,
ripe and gold as grain fields,
Black Bear smells that it is time to nest,
to dig her winter den,
slipping inside its shadowy warmth—
time again to hibernate.

All through the night, Black Bear dreams
of the dark perfumes of underground roots
and the deep scents of earth slumber.
She remembers the fragrance of newborn cubs.

As hibernation approaches, an adult female black bear's diet may expand to as many as 20,000 calories—the equivalent of about 150 scoops of vanilla ice cream—each autumn day during a period of intense eating, known as **hyperphagia**.

When food becomes scarce and the weather grows cold, she can find a den under thick tree roots, inside a hollow stump, or within an open cave. Otherwise, she can make one by digging an underground hole with her front claws.

FACT:

Dens tend to be cozy and are often little bigger than a black bear's body.

During hibernation, an adult female black bear does not eat, drink, **urinate**, or **defecate**. Because her body remains clean, a **predator** hunting for her—for example, a cougar—cannot smell her den.

While she hibernates, the rhythms of her body slow down. She breathes half as frequently, and her heart beats at one quarter its normal speed. Her body temperature also drops several degrees Fahrenheit.

Temperature In Farenheit:
Waking 100 degrees
Hibernating 88 degrees

Beats Per Minute:
Waking Heart Rate 80-100
Hibernating less than 22

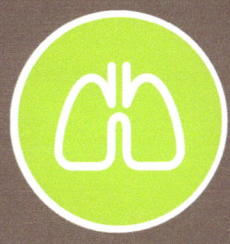

Breaths Per Minute:
Hibernating 1 to 2

Nevertheless, if the adult female black bear is pregnant, between one and five babies will grow within her belly to be born in the winter.

Despite her body's changes during hibernation, if a mother black bear is threatened in her den, she will revive quickly to defend herself and her cubs once they are born.

WINTER

When the winter moon rises,
cold and white as snowfall,
Black Bear feels that it is time to change,
to give birth to her cubs,
becoming a source of precious life—
time to renew the world.

All through the night, Black Bear dreams
of the Great Mother Bear constellation
and the star cub curled beside her.
She remembers furry clouds shifting their shapes.

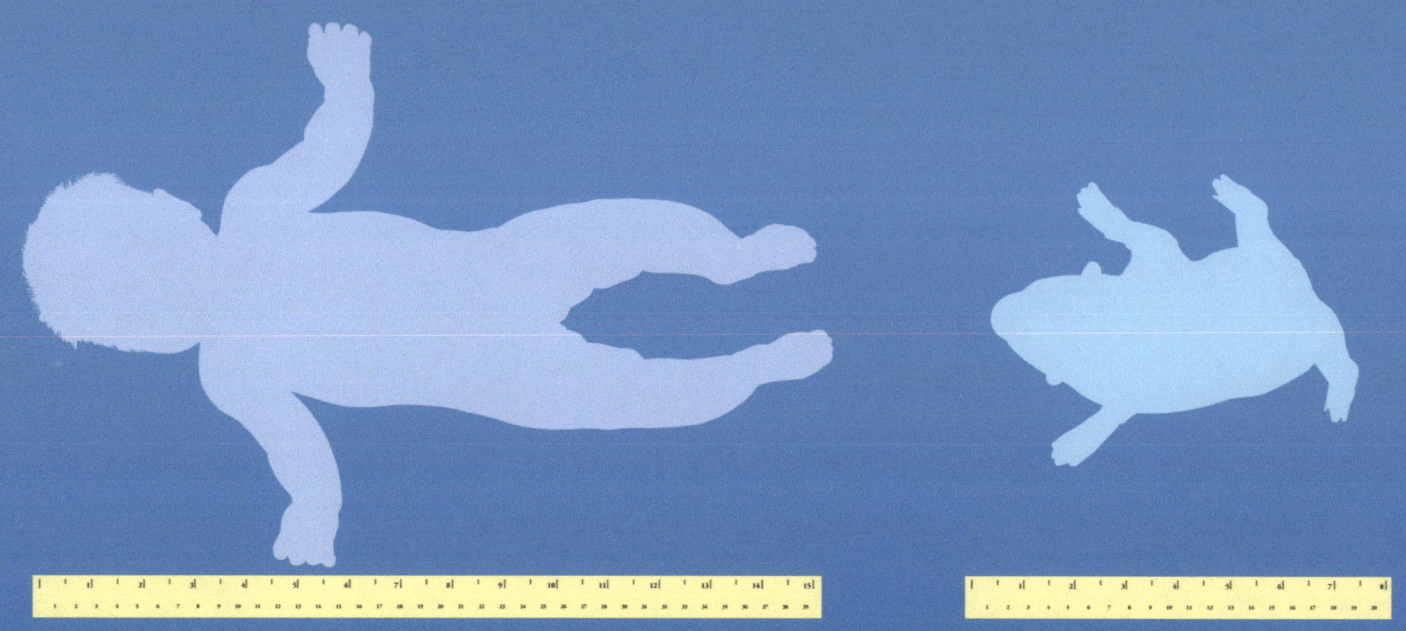

7.5 pounds (3.5 kilograms)
20 inches (50 centimeters)

.75 pounds (.34 kilograms)
8 inches (20 centimeters)

In January, a mother black bear delivers her cubs in the den. On average, each **litter** includes two babies. Each newborn cub typically weighs between one half and three quarters of a pound—about one tenth the average weight of a human baby. Newborn black bear cubs also tend to be half the length of human infants.

At birth, black bear cubs have almost no fur. They sleep against their mother's belly to stay warm and to drink her milk.

FACT: Newborn cubs will not open their eyes until they are about six weeks old.

When spring arrives, the cubs leave the den with their mother, having grown to weigh between four and six pounds each.

In contrast, during the months of hibernation, a nursing mother black bear loses about one third of her body weight.

BEFORE HIBERNATION

AFTER HIBERNATION

FACT: In the den, mother black bears lose almost ten times as much weight as newborn cubs gain.

A NEW YEAR

When the first spring moon returns,
fresh and free as March winds,
Black Bear sees that it is time to lead,
to show her cubs the world,
teaching them her wilderness wisdom—
time to raise clever bears.

All through the night, Black Bear dreams of the flowering and fading of spring, and the ebb and flow of rainfall. She remembers viewing the vast horizon.

Only the mother black bear cares for her cubs. The little bears may never meet their father or grandparents.

After hibernation, the mother black bear lets the cubs drink her milk, as they did in the den. However, when their first teeth arrive, she shows the cubs how to forage for solid food.

FACT:

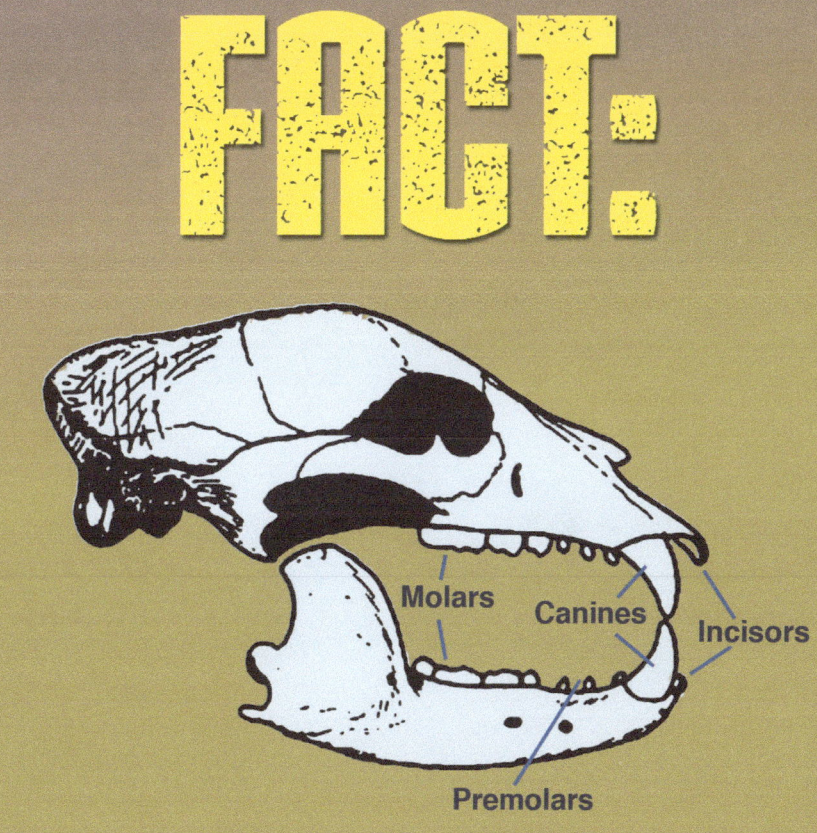

Black bears have 42 sharp teeth: 12 incisors for snipping forest greens and flowers, 4 fang-like canines for tearing open fallen logs ripe with insects, and 16 premolars and 10 molars for grinding seeds and meat.

Preparing the cubs for life without her, the mother black bear teaches them how to run, to swim, and to climb trees.

When autumn comes, the mother black bear builds a new den for herself and her cubs, gathering fallen leaves and twigs to cover the dirt floor of their retreat.

FACT:

The black bear family will hibernate in their den during the winter, then rise again the following spring.

Throughout their year and a half together, the mother black bear protects her cubs from danger. If the cubs stray, she will grunt at them until they follow her again. If a possible predator, such as a grizzly bear, a cougar, or a human, approaches the cubs, the mother black bear may roar and charge at them, sometimes even attacking the predator.

However, by their second June as a family, the mother black bear will encourage the cubs to make their own way in the wild.

If you encounter a black bear:
Stand tall, and raise your hands to appear bigger to the nearsighted bear.

Do not look into the black bear's eyes. Such eye contact can be mistaken for a threat.

Slowly back away from the black bear until you have reached safety.

Do not run or climb a tree because the black bear may chase after you or follow you up into the branches.

If the black bear walks toward you, clap your hands and make noise to scare the bear away.

Do not say or shout the word "bear." Other people may have accidentally taught the black bear that this word means, "Come here for some food to eat." **Never feed a bear.**

THE FIVE SENSES OF BLACK BEARS

In this book, you have read about black bears' five senses: hearing, taste, smell, touch, and sight. By far, black bears' sense of smell is their most powerful: **The nose knows!**

FACT: Black bears can smell honeycomb from a mile away, roughly the distance of 2,000 skateboards lined up end-to-end.

Black bears also have highly sensitive ears. They can hear people talking from as far away as the length of a soccer field.

"HEY!"

By comparison, black bears' sight is less impressive. They see in color and have excellent close-range vision for finding safe food to eat. However, at distances greater than 30 yards—about the length of a school swimming pool—black bears do not see details well.

In addition, black bears have short, strong claws for gripping tree trunks as they climb. The ten toenails on their front paws can also do the fine work of pulling bushes and tree branches to the black bears' open lips and long, pink tongues.

FACT:

Black bears live for 15-25 years, on average.

In every season of their lives, black bears use their five senses to survive, enriching the world as they wander.

QUESTIONS FOR DISCUSSION

Would you like to hibernate? Why or why not?

When celebrating with your family or friends, have you ever eaten a lot of food like a black bear during hyperphagia? If so, what happened?

What would you do if you met a black bear cub in the woods? Why?

What is your favorite thing that you have learned from your mother or another adult? Why?

ABOUT THE AUTHOR

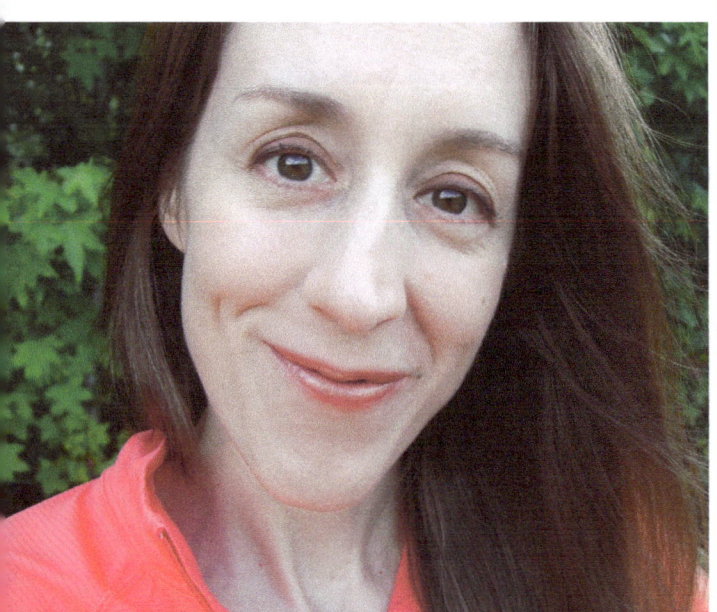

Anne Michelle Lawrence lives in Florida, where black bears of the subspecies *Ursus americanus floridanus* are her native neighbors. She loves to hike with her daughter.

GLOSSARY

carrion (KAR-ee-uhn), *noun*: the meat of dead animals

defecate (DEF-uh-kayt), *verb*: to poop

diurnal (dye-UR-nuhl), *adjective*: awake and active during the day

forage (FOR-ij), *verb*: to search for food

hibernate (HYE-bur-nate), *verb*: when an animal's body processes slow down and they take a long rest during the cold months when food is scarce

hyperphagia (hye-pur-FAY-zhuh), *noun*: intense eating in preparation for a long fast, or time without food, during hibernation

litter (LIT-ur), *noun*: the number of animal babies born at one time to the same mother

nocturnal (nahk-TUR-nuhl), *adjective*: awake and active during the night

omnivores (AHM-nuh-*vorz*), *noun*: eaters of both meat and plants

predator (PRED-uh-tur), *noun*: an animal that kills and eats other animals

urinate (YOOR-uh-*nate*), *verb*: to pee

www.ingramcontent.com/pod-product-compliance
Lightning Source LLC
Chambersburg PA
CBHW050856010526

44118CB00005BA/178